Contents

SCOTTISH DEVELOPMENT DEPARTMENT

THE NEW

BULLETIN

METRIC SPACE STANDARDS

WITHDRAWN

EDINBURGH

HER MAJESTY'S STATIONERY OFFICE 1968

First published, 1968
Fifth impression, 1985

ISBN 0 11 490148 1

1
Scope of the Bulletin

1.1   This is the first of a series of publications, dealing with the design and layout of housing developments in the public sector, which will update and replace the Scottish Housing Handbook.

1.2   The present publication gives general advice on internal design standards for dwellings. It is essentially an introductory document, dealing principally with the standards of space, fittings, and equipment, which will apply to developments designed in metric dimensions, after 1st January, 1969. It therefore replaces the imperial standards set out in the Scottish Housing Handbook Part 3—'House Design', 1956 edition.

1.3   This document should be regarded therefore with these limited objectives in mind. It is not, on its own, a complete primer for the design of dwellings. It introduces a new design philosophy, sets out the design framework required, and provides the necessary information on the detailed application of the new standards. The material is set out so that standards are stated on one page, with a commentary on the standards placed opposite, explaining their application. Illustrations are included where appropriate.

1.4   The standards themselves result from co-operation between the Department and housing authorities of the Scottish Local Authorities Special Housing Group. Future publications in the series, dealing with other design standards which influence the dwelling and its relationship to housing layout, will be based on the results of further joint research, through a continuing series of user requirement studies, and on similar research studies elsewhere

2

Introduction: The New Standards

2.1 **METRICATION OF HOUSEBUILDING**

As a result of the Government decision to support the change to the metric system, proposed by the Federation of British Industries*, the British Standards Institution published the Programme for the Change to the Metric System in the Construction Industry, (PD 6030), in February, 1967. The implications of this programme for housing authorities in Scotland were outlined in SDD Circular 27/1968, Metrication and Dimensional Co-ordination in Public Sector Housing, which indicated that revised advice on housing standards would be published in metric, by the Department, in good time to allow designers to change to the production of metric drawings and documents in accordance with the programme. This Bulletin sets out the new metric standards.

2.2 **REVIEW OF STANDARDS**

Since 1956 space and accommodation standards for public authority housing in Scotland have been those set out in Part 3 of the Scottish Housing Handbook, 'House Design'. While these standards have generally proved to be adequate in terms of overall floor space, they have now been in use for over a decade, and some of the recommendations have been overtaken by changing social needs, and the mass of new information on all aspects of house design and layout, which has emerged from research studies. Accordingly the Department undertook an intensive review in order to establish what new standards were desirable.

The need for updating of design information was reinforced by the change to metric and the associated move towards dimensional co-ordination. The systematic changeover by designers and manufacturers to a metric dimensional framework established the need to study and decide what the new metric sizes of components should be to provide the best industrial and design advantages. At the same time an opportunity was created to rationalise and co-ordinate the new sizes, so that components of all kinds might fit together with the minimum waste of effort or material. A cardinal principle of such rationalisation is the elimination of non-essential variations in consumer demand, and it became clear that if the full benefits of rationalisation were to be felt throughout the United Kingdom, it would be desirable to have a single set of housing standards for the whole country.

The process of reviewing and updating of standards undertaken by the Department immediately revealed a similarity between the standards desirable in Scotland and the metric standards which are being introduced in England and Wales, on 1st January, 1969. So similar were the standards that an opportunity existed for unification and the formation over a period of time of common standards of space and accommodation for housing. There are many advantages to be won from such a step in terms of increased efficiency, the cost benefits of standardisation, the elimination of duplication of work and the pooling of research and development in the professions and in the industry. It becomes much more feasible to plan for large scale housing programmes, using industrialised and rationalised-traditional building methods, based on widespread use of dimensionally co-ordinated components.

* now incorporated in the Confederation of British Industry

For these reasons, therefore, the new space and heating standards introduced in this document are the same as those being introduced in England and Wales, which in turn incorporate the principal recommendations of the Parker Morris Report 'Homes for Today and Tomorrow'.

2.3   THE NEW STANDARDS

At every stage in the life of a family, the home has to provide for an extremely wide range of activities. It has to accommodate different individual and group interests and activities, involving any number or all of the family, while providing reasonable privacy for the individual or group. The normal process of family development means that these requirements are constantly changing, and demand arises for an adaptable home, which can respond to changing social needs.

It is difficult to meet this need for flexibility, using standards which are framed in terms of minimum room sizes, since emphasis on room size tends to focus the designer's attention on working out a pattern of room areas, which comply with the standards, and assumes a conventional arrangement of the dwelling, and the particular way in which a given room will be used.

Accordingly, the new standards are based on minimum overall floor areas, which reflect varying sizes of household, and compliance with mandatory minimum room areas, as such, is no longer required. It is assumed instead that the usefulness of a room depends on whether its shape, and the position of doors and windows allow an appropriate scale and arrangement of furniture. The approach is to define what activities are likely to take place, to assess the furniture and equipment necessary, and then to design round these needs.

As a minimum, each room must be capable of containing certain specified items of furniture or equipment, arranged in a sensible way, with sufficient space left to make the room comfortable and convenient in use. The plan relationship of one room with another is the result, rather than the starting point of the design process, and rooms themselves grow from the needs of the occupier, evolving as a consequence of thought, and not simply through copying what has gone before.

Given the minimum area for the whole dwelling, the designer is free to arrive at the best way to arrange the space and equipment to meet the requirements of the particular size of family, within the limitations and opportunities provided by the site, the structural possibilities and the cost.

Thus the emphasis of the new standards is placed on the activities that people want to pursue in their homes. The design process starts with a recognition of these various activities, and their relative importance in social, family and individual lives, and goes on to assess the conditions necessary for their pursuit, in terms of space, atmosphere, efficiency, comfort, furniture and equipment, organising together those activities that demand it, separating those which cannot be carried on together or near one another, considering frequency, time and sequence of the activities, as well as place.

Under the new standards the designer will have greater freedom to design homes which will meet current and future needs. He can use this freedom to ensure flexibility in plan arrangement, giving the occupier options for change. The more flexible the internal space, the greater is the likelihood that the dwelling will be able to satisfy the demands made on it by a family at different stages of its development, by new occupants, by a rising standard of living and by changing concepts of good design. The standards demand imaginative planning and allow for the achievement, within indicative costs, of high quality design as well as the satisfaction of basic housing needs.

2.4	**APPLICATION OF THE NEW STANDARDS**	The new standards are applicable only to dwellings designed in metric dimensions from 1st January, 1969, and will be mandatory for all public authority housing developments submitted for subsidy or borrowing consent on or after 1st January, 1972.

Plans prepared to the new standards should take account of B.S.4330: 1968, 'Recommendations for the Co-ordination of Dimensions in Building—Controlling Dimensions', which incorporates the dimensional framework of the D.C. Series of publications by the Ministry of Public Building and Works, and the recommendations contained in Design Bulletin 16—'Co-ordination of Components in Housing—Metric Dimensional Framework' published by the Ministry of Housing and Local Government in consultation with the Department. As indicated in SDD Circular 27/1968, public authority submissions which depart unreasonably from these recommendations may not be approved during the transitional period and compliance with the recommendations will be mandatory, in the public sector, from 1st January, 1972.

2.5	**NEW STANDARDS AND THE BUILDING REGULATIONS**	Dwellings designed to the new space standards may not in all cases comply with the minimum areas specified in Part XV of the Building Standards (Scotland) Regulations. Amending regulations giving effect to the new standards are under consideration. In the meantime where relaxation of the regulations is required to enable dwellings to be built to the new standards, the procedure under Section 4 of the Building (Scotland) Act, 1959 will be available.

2.6	**NEW STANDARDS AND INDICATIVE COSTS**	A cost conscious approach to housing design is essential and is emphasised in the commentary to the new standards. To ensure that consideration of costs is taken fully into account from the earliest planning stages a system of cost planning and indicative costs for local authority housing has been introduced, and will be fully effective from 1st January, 1969. The basic principle of the system is that by means of published tables of indicative costs, authorities will know, from the outset of the planning of their housing proposals, the level of cost which will be accepted for the purpose of subsidy.

7

The published tables show the costs for which dwellings of given sizes and built at given densities can reasonably be provided. The costs do not relate to a dwelling of given shape, construction or specification: they represent instead the upper limit of expenditure which it is reasonable to incur to meet a given housing requirement, having regard to the standards to which housing is built, the density of the development and the average size of the families accommodated.

Within this upper limit of expenditure the designer has a considerable degree of freedom to decide on choice of types, plan forms, construction and specification. It is essential, however, that the scheme as a whole is fully cost planned, and that the elements of cost for each dwelling type of his choice are clearly identified.

3
Definitions

3.1   The following definitions are relevant only to this Bulletin and should not be confused with those contained in the Building Regulations or the Housing Acts.

3.1.1   DWELLING: Any place of residence, including a house, a flat or a maisonette.

3.1.2   HOUSE: A dwelling divided vertically from every other dwelling and with its principal access from ground level.

3.1.3   FLAT: A dwelling on one floor, forming part of a building from some other part of which it is divided horizontally.

3.1.4   MAISONETTE: A dwelling on more than one floor, forming part of a building from some other part of which it is divided horizontally.

3.1.5   SINGLE ACCESS HOUSE: A 'house' with public access from one side only.

3.1.6   DUAL ACCESS HOUSE: A 'house' with public access from both sides.

3.1.7   WATER CLOSET: The term refers to the compartment and not the fitting.

3.1.8   BUILDING REGULATIONS: The Building Standards (Scotland) Regulations 1963 and Amendments.

3.1.9   METRIC ABBREVIATIONS: m : metres
mm : millimetres
m^2 : square metres
m^3 : cubic metres

NOTE ON METRIC DIMENSIONS   Linear dimensions on all diagrams are given in millimetres.
Where dimensions shown are the subject of a British Standard, the metric sizes given are intended as reasonable guides only, and will be superseded by the metric standards when these are published.

4.1

Space Standards

4.1 Dwelling Areas

A dwelling for occupation by the number of people shown in the table below must be designed to provide areas of net space and general storage space not less than those set out in the table and fulfilling the conditions in Section 4.2.

N=net space (see 4.2.1) S=general storage space (see 4.2.2)		Number of people (i.e. bed-spaces) per dwelling						
		1	*2*	*3*	*4*	*5*	*6*	*7*
		m²	*m²*	*m²*	*m²*	*m²*	*m²*	*m²*
Houses								
1 storey	N	*30*	*44·5*	*57*	*67*	*75·5*	*84*	
	S	*3*	*4*	*4*	*4·5*	*4·5*	*4·5*	
2 storey (semi or end)	N				*72*	*82*	*92·5*	*108*
	S				*4·5*	*4·5*	*4·5*	*6·5*
(intermediate terrace)	N				*74·5*	*85*	*92·5*	*108*
	S				*4·5*	*4·5*	*4·5*	*6·5*
3 storey	N					*94*	*98*	*112*
	S					*4·5*	*4·5*	*6·5*
Flats	N	*30*	*44·5*	*57*	*70**	*79*	*86·5*	
	S	*2·5*	*3*	*3*	*3·5*	*3·5*	*3·5*	
Maisonettes	N				*72*	*82*	*92·5*	*108*
	S				*3·5*	*3·5*	*3·5*	*3·5*

*(67 if balcony access)

Tolerance: Where dwellings are designed on a planning grid and not otherwise, a maximum minus tolerance of $1\frac{1}{2}\%$ will be permitted on the Net Space.

4.2   Rules of Measurement

4.2.1

Net space *is the area on all the floors enclosed by the walls of a dwelling measured to unfinished faces.*

It includes: *the space, on plan, taken up on each floor by*
any staircase,
partitions
any chimney breast, flue and heating appliance
the area of any external W.C.

It excludes: *the floor area of general storage space (S in table)*
dustbin store,
fuel store,
garage
balcony
any area in rooms with sloping ceilings to the extent that the height of the ceiling does not exceed 1·5 metres,
any porch, lobby or covered way open to the air.
In the case of a single access house any space within a store required to serve as access (taken as 700 mm wide) from one side of the house to the other must be provided in addition to the floor areas in the table.

4.2.2

General Storage Space *must have a minimum floor to ceiling height of 1·5 metres.*

It excludes: *fuel storage (Section 4.4.2),*
linen storage (Section 4.4.3),
kitchen storage (Section 6.2),
any dustbin store,
pram space located in a store,
in the case of a single access house, any space within a store required to serve as access (taken as 700 mm wide) from one side of the house to the other.

4.3   Floor to Floor Height

For public authority schemes submitted for subsidy or borrowing consent on or after 1st January, 1972, a floor to floor height of 2·6 metres will be mandatory.

For public authority schemes submitted for subsidy or borrowing consent before that date, it is recommended that wherever possible authorities should adopt a floor to floor height of 2·6 metres.

In the case of the top storey, a dimension should be chosen to suit the use of standard components, and the clear floor to ceiling height should then normally be the same as the floor to ceiling heights of the lower storey or storeys.

COMMENTARY

4.2.1   The diagram below illustrates which storage components are included, and which are excluded, from the net space areas set out in the table.

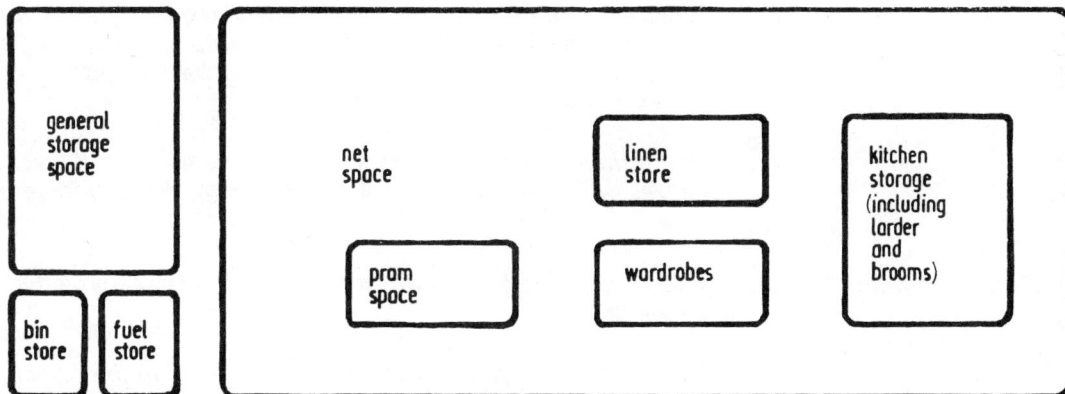

4.2.2

The space below a staircase may be used as General Storage Space, where it exceeds a height of 1·5 m, provided that the Net Space Area is maintained.

It may also be used as pram space or additional storage.

4.4
Storage

4.4 Storage Space

4.4.1 GENERAL STORAGE

(a) Houses

Some of the storage space may be on an upper floor, but at least 2·5 m² must be at ground level.

When some of the storage space is provided on an upper floor, it must be enclosed separately from linen or bedroom cupboards; it must be accessible from the circulation space, or from a room if conveniently accessible in relation to furnishing.

When there is a garage integral with or adjoining a house, any area in excess of 12·0 m² may count towards the general storage provision.

(b) Flats and Maisonettes

Not more than 1·5 m² of the prescribed area may be provided outside the dwelling.

Where there is a garage integral with or adjoining the dwelling, any area in excess of 12·0 m² may count towards this 1·5 m².

NOTE: General storage space must be separately enclosed and floored, as required by the Building Regulations.

4.4.2 SOLID FUEL STORAGE (where applicable)

(a) Houses

1·5 m² where there is only one appliance.

2·0 m² where there are two appliances, or in rural areas.

(b) Flats and Maisonettes

1·0 m².

Fuel storage provision must comply with the requirements of the Building Regulations.

4.4.3 LINEN STORAGE

All Dwellings

A cupboard must be provided giving the following storage provisions:

4 person and larger dwellings:	0·6 m³
1/2/3 person dwellings	0·4 m³

4.4.4 KITCHEN STORAGE

See Section 6.2.1.

4.4.5 PRAM SPACE

See Section 7.2.2.

4.4.6 MEDICINE CUPBOARDS

COMMENTARY

4.4.1 The positioning and accessibility of storage space is at least as important as the volume available. It is desirable therefore that general storage space be distributed about the dwelling, and there may be a positive advantage in not having it all in one place.

In houses, some general storage may be accessible from outside, and in flats and maisonettes the designer also has the option of providing part of the mandatory requirement outside the dwelling, for storing large toys, bicycles, etc. Where external storage is provided, precautions must be taken to avoid dampness and condensation within the store.

4.4.2 Fuel storage should normally be accessible under cover from the dwelling and it should not be necessary for deliveries to be made through the dwelling.

4.4.3 Since a linen cupboard must be perfectly dry it may not be desirable to place it on an outside wall or in the bathroom or kitchen. It may be combined with a suitable hot water cylinder. It should have slatted shelves and be conveniently shaped. Where the hot water cylinder is in the linen store the mandatory cubic capacity must be in addition to the space occupied by the cylinder and other fittings.

4.4.6 In addition to other storage, it is desirable to provide a safe medicine cupboard in every new dwelling. Such a cupboard should conform to British Standard 3922/67 and should be placed, preferably in the bathroom, at a height of not less than 1·4 metres above finished floor level.

5
Space Heating

5.1   All Dwellings

The minimum standard must be an installation with appliances capable of maintaining kitchen and circulation spaces at 13°C, and the living and dining areas at 18°C when the outside temperature is minus 1°C.

COMMENTARY

5.1 Trends

People increasingly expect to feel warm in their homes, and it is therefore not surprising that partial or whole house heating has become a feature of the design of dwellings in recent years.

Bedrooms

While the minimum standard does not require bedroom heating it will often be possible because of dwelling design and type of installation, to provide heating to bedrooms at very little extra cost. It is recommended that this be done where it can be achieved within the indicative cost limits.

Running and maintenance costs

The advantages of a heating system are lost if the occupier finds it too expensive to use. It is essential therefore that the dwelling and the heating system are designed having regard to the running costs of the installation. The costs for water heating and maintenance must also be considered. These costs are extremely difficult to predict, and it is only through establishing records of cost in use that realistic figures are available for comparison.

Thermal insulation

Better thermal insulation reduces heat loss and results in lower running costs, whilst adding to the thermal comfort of the occupier. Air temperature alone does not guarantee thermal comfort, since the body loses heat by radiation to surrounding cold surfaces in a room. By permitting higher internal surface temperatures, correctly placed thermal insulation reduces discomfort from this source and reduces the incidence of condensation. Wherever possible, within indicative costs, designers should aim to provide a better standard of thermal insulation than the minimum requirements of the Building Regulations.

Plan arrangement

Plan arrangement plays an important part in the satisfactory heating of a dwelling. Extended plan arrangements which result in a bedroom being isolated from the heated areas may give rise to condensation problems.

Internal planning

Rooms should be planned to avoid draughts and it should be possible to place furniture so that occupants are not compelled to sit next to large areas of single glazed window or door openings. Consideration should be given to draught stripping external windows and doors.

Windows

Large windows result in areas of low surface temperature, with substantial heat loss and consequent discomfort to occupants. Windows should generally be no larger than necessary to meet daylighting and sunlighting requirements.

Ventilation

Ventilation is closely related to thermal comfort and requires special attention when using heating methods which dispense with flues. The openable areas of windows should be so designed that the degree of ventilation can be finely controlled, particularly in spaces where condensation originates such as kitchens and bathrooms. Standards for ventilation are laid down in Part X of the Building Regulations.

Condensation

Severe problems of condensation have become more frequent in recent years. The subject has been studied by the Department in collaboration with the Ministry of Public Building and Works, and designers are referred to the resulting Bulletin, 'Condensation in Dwellings: Pt 1, A Design Guide'*, which gives advice on how this problem may be minimised.

* Published by HMSO

6.1

Fittings and Equipment

6.1 Bathrooms and W.C.s

6.1.1	**ALL DWELLINGS EXCEPT THOSE LISTED IN PARA: 6.1.2**	*1 W.C. fitting, which may be in the bathroom.*
6.1.2	**2 AND 3 STOREY DWELLINGS AT OR ABOVE THE MINIMUM FLOOR AREA FOR 5 PERSONS; SINGLE-STOREY DWELLINGS (INCLUDING FLATS) AT OR ABOVE THE MINIMUM FLOOR AREA FOR 6 PERSONS**	*2 W.C. fittings, one of which may be in the bathroom.*
6.1.3	**OTHER PROVISIONS**	*All other provisions must be in accordance with the requirements of the Building Regulations.*

COMMENTARY

6.1   The following additional standard should be achieved wherever possible:

2- and 3-Storey Dwellings   1 W.C. fitting, which must be in a separate compartment.
for 4 persons; Single-
Storey Dwellings (including
Flats) for 4/5 persons:

6.1.2   A second W.C. should be so placed that it can be used conveniently by guests and by children coming in from the garden.

6.1.4   **General**

Attention is drawn to the following aspects of the planning of bathrooms and W.C.s:

(a) The space needs of different activities: e.g., bathing and drying a baby, washing the hair, using the W.C. (diagrams 29-32).

(b) The positioning of fittings: e.g., to simplify cleaning, to permit access to a window without having to stand on or in a fitting, to suit the physical capabilities of children and the elderly.

(c) Possible future needs: where cost permits, bathrooms should not be so tightly planned as to rule out additional equipment such as bidets and showers. Changing family needs should also be considered, e.g., it may become desirable, as the family grows, to sub-divide the bathroom so that the W.C. fitting is enclosed separately.

6.2
Fittings and Equipment STANDARDS

6.2   Kitchens

6.2.1   STORAGE

Kitchen fitments comprising enclosed storage space in connection with

(a) preparation and serving of food and washing up,

(b) cleaning and laundry operations, and

(c) food storage,

must be provided as follows:—

3 person and larger dwellings	*2·3 m³*
1 and 2 person dwellings	*1·7 m³*

This provision must include:

(i) a ventilated larder of not less than 0·17 m³

(ii) a broom cupboard which may be provided elsewhere than in the kitchen.

The rules of measurement for the cubic capacity of kitchen storage: fitments are those contained in Schedule 1 to the Building Regulations.

6.2.2   REFRIGERATOR AND WASHING MACHINE

In addition to kitchen storage, the sink and space for a cooker, a minimum of two further spaces must be provided in convenient positions to accommodate a refrigerator and washing machine. The washing machine may be in a kitchen or in a convenient position elsewhere, e.g., in a utility room. Both spaces may be provided under worktop surfaces.

6.2.3   LAUNDRY AND DRYING FACILITIES

Provision must be made in accordance with the Building Regulations.

6.2.4   CASUAL MEALS

Any working kitchen in a dwelling for 2 or more persons must provide a space where casual meals may be taken by a minimum of 2 persons. (See Section 7.3.3.)

COMMENTARY

6.2.1   It should be noted that kitchen storage provision forms part of NET SPACE and is separate from GENERAL STORAGE SPACE (see Section 4.2.1. Commentary).

6.2.2   Sufficient space should be allowed for the kinds of washing machines which are in current use : sizes vary from 450-600 mm in depth to 550-800 mm in width and 750-950 mm in height (see diagrams 77 and 80). Wherever possible, the space needs of equipment such as mixers and dishwashers, which may be provided by the tenant during the lifetime of the dwelling, should be allowed for at the planning stage.

6.2.4   Space for casual dining should be provided clear of any traffic way and working area and well away from cookers. (See Section 7.3.2. Commentary and Diagram 13.)

6.2.5   Kitchen Planning

Particular attention should be given to the following aspects of kitchen planning, in addition to the above :

(a) Worktops should be provided on both sides of the sink and on both sides of the cooker position. Fitments should be arranged to provide a work sequence of WORKTOP—COOKER—WORKTOP—SINK —WORKTOP (or the same in reverse order) unbroken by a door or other traffic way. (See diagrams 1-2.)

(b) The cooker should not be immediately next to a window or door and worktops on each side of a cooker should be at the same height as the boiling plate or rings (diagrams 3 and 11).

(c) Good, shadowless lighting over the cooker and working surfaces is essential, both during the daytime (see Building Regulation 132) and after dark.

(d) Equally important is the provision in the kitchen and laundry areas of controllable ventilation, to minimise condensation (see Section 5) and the spread of smells to other parts of the house. (See Regulation 114.)

(e) It should be possible to reach window catches and storage space without having to stand on chairs or stools. (See diagrams 5-8.)

(f) Doors to the kitchen should be placed to minimise nuisance and danger from through traffic. The swings of all doors and cupboard doors should be planned to avoid collision (see diagrams 11 and 12), and sliding doors are preferable wherever the Building Regulations permit.

GENERAL
Further research is being undertaken into the problems of kitchen planning and a new British Standard is in the course of preparation.

6.3
Fittings and Equipment STANDARD

6.3 **Power Points**

6.3.1 **ALL DWELLINGS** *Provision of power points must be in accordance with the Building Regulations.*

COMMENTARY

6.3.1   Attention is drawn to Section 6.2.2 which requires space for a refrigerator and washing machine. Additional power points are therefore required in accordance with Building Regulation 195 (3).

It is important to position socket outlets where they are needed so that long trailing flexes are avoided, as far as possible. (See Commentary 7.3.1(e).)

Electric towel rails, or other appliances having heating elements, which could be touched by someone using a bath or shower, are not permitted under the Building Regulations.

Meters and fuse boards should be placed so that they are readily accessible and preferably high enough to be out of reach of small children. (See Commentary 7.2(e).)

Position and heights of power points in kitchens should be carefully considered in relation to kitchen layout and heights of working surfaces.

7

Planning the Dwelling

7.1 Access to Dwellings

7.1.1 HOUSES

Every single-access terraced house designed for three or more persons must have a way through the house from front to back, and this must not be through the living room. In such cases it should not be necessary to take the dustbin through the house for collection.

7.1.2 FLATS AND MAISONETTES

The standard with regard to maximum walk-up height and provision of lifts is set out in Building Regulation 182.

7.2 Entrance and Circulation:

7.2.1 HALL OR LOBBY

The entrance hall or lobby in each dwelling must be provided with space for hanging outdoor clothes. (See diagrams 37-38.)

7.2.2 PRAM SPACE

In three person or larger dwellings at ground level or served by a lift or ramp, space must be provided for a pram (1 400 × 700 mm).

COMMENTARY

7.1.1   Single access terraced house; the following diagrams show some methods of providing the required through route:

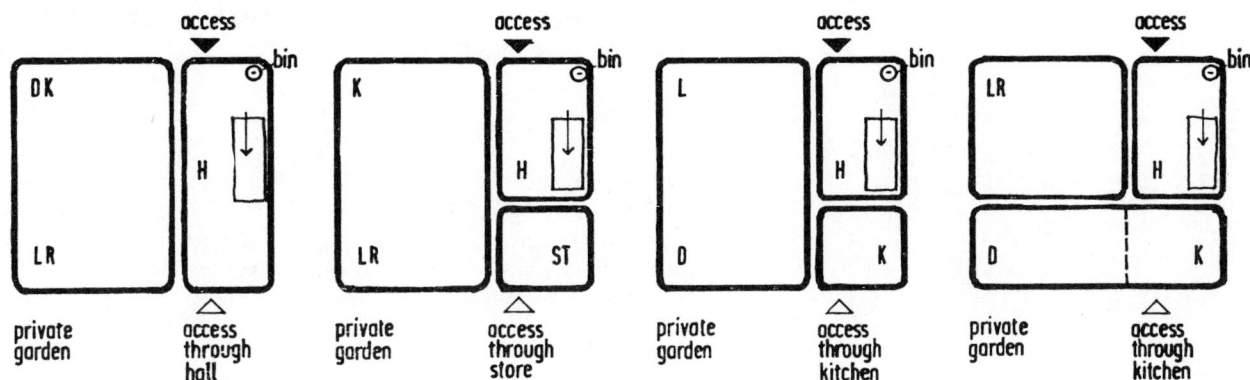

7.1.2   For families with young children, the highest floor for daily living purposes should not be more than two storeys from the ground.

7.2   In the planning of entrances, stairways and general circulation areas, it should normally be possible:

(a) To reach bedrooms from the entrance without having to enter the main living space.

(b) To reach the main living space from the main entrance without entering the kitchen.

(c) To reach the bathroom and a W.C. from the bedrooms without entering any other room.

(d) To move round the dwelling large items of furniture and equipment. (See diagrams 41 and 42.)

(e) To read meters from outside the dwelling, or at least without having to enter the living spaces.

(f) To avoid the use of winders at tops of stairs, recessed and projecting treads, and risers in unexpected positions.

7.2.1   The hall or lobby should be:

(a) Big enough to receive visitors.

(b) Big enough to allow a pram and furniture to be brought indoors. (See diagram 40.)

(c) So arranged as to prevent callers intruding on the privacy of the living areas.

7.2.2   Wherever possible, pram spaces should be provided in ALL dwellings designed for 3 persons or more.

It should be possible:

(a) To store a pram at or near an entrance without entering the living spaces and without obstructing circulation areas.

(b) To move the pram around the dwelling without undue difficulty.

25

NOTE TO PARAS.: 7.3 AND 7.4:   ALL DWELLING PLANS MUST SHOW THE FURNITURE AND KITCHEN EQUIPMENT DRAWN ON AND SHOULD BE DESIGNED TO ACCOMMODATE THE FURNITURE SET OUT IN PARAS. 7.3 AND 7.4 BELOW. (FOR DIMENSIONAL DATA, SEE DIAGRAMS 1-97 ON PAGES 31-43.)

7.3   Living Areas

7.3.1   LIVING SPACE

Furniture: 2 or 3 easy chairs
a settee
a television set
small tables
a reasonable quantity of other possessions such as:
radiogram
bookcase

7.3.2   MEALS SPACE

Furniture: a dining table and chairs

7.3.3   KITCHEN

Furniture: a small table unless one is built in

Detailed standards for the planning of kitchens are set out in Section 6.2.

COMMENTARY

7.3.1 In planning the arrangement of furniture in the living space, it must be remembered that the home needs to make provision both for the family to gather together and for its members to carry on their own pursuits.

At the planning stage, therefore, the following points are most important:

(a) The main living space should be large enough to accommodate the furniture necessary for the whole family and occasional visitors, and to provide for alternative furniture arrangements.

(b) It should be possible to shut off the main living space from the rest of the dwelling.

(c) In dwellings for 4 persons and more a second living space is desirable which can cater for activities needing privacy and freedom from disturbance.

(d) Doors and windows should be positioned to minimise draughts and to leave as much wall space as possible for furnishing.

(e) Radiators, warm air outlets, television and power points should be positioned to make furnishing as flexible as possible.

Diagrams 16-19 illustrate typical arrangements of furniture for various activities.

7.3.2 Three principal combinations of meals spaces and kitchens are possible:

(a) The working kitchen enlarged to provide meals space for the whole family, there being no other meals space. This does not allow separation of working from eating areas.

(b) The working kitchen with a separate meals space, adjacent to and with access from the working area.

(c) A separate room, which can be used for meals with all the family present, but which is available for other activities needing privacy.

Whichever arrangement is chosen, the main meals space should always be large enough for the whole family and occasional visitors.

In larger homes the meals space should be separate from the kitchen and main living space, and should be able to accommodate a sideboard and easy chair as well as the dining table and appropriate number of chairs.

In dwellings for two persons and more, table or counter space must be provided in the kitchen, where casual meals can be taken. (See Section 6.2.4 and diagram 13.)

Wherever possible, the working area of the kitchen should be screened from the main dining space.

Diagrams 13-15 illustrate space requirements for dining areas.

Note that minimum aggregate areas for living/dining and dining/kitchen areas are no longer required.

7.4 Bedrooms

7.4.1 DOUBLE BEDROOMS

*Furniture: A double bed (2 000 × 1 500 mm) or
2 single beds (2 000 × 900 mm)
bedside tables
chest of drawers
a double wardrobe **or**
space for a cupboard to be built in
dressing table*

7.4.2 SINGLE BEDROOMS

*Furniture: bed or divan (2 000 × 900 mm)
bedside table
chest of drawers
a wardrobe **or**
space for a cupboard to be built in*

7.4.3 STUDY BEDROOMS

*Furniture: as for single bedrooms plus
desk
chair
bookcase*

7.4.4 BEDSITTER BEDROOMS

Furniture: as for study bedrooms plus at least one easy chair.

7.4.5 GENERAL

At least one double-bedroom should be capable of containing a double bed.

When single beds are shown in a double bedroom they may abut. If alongside walls, however, a space of at least 750 mm must be left between them. (Diagrams 25-26.)

The wardrobe or cupboard for a second double bedroom may be provided within easy access outside the room.

Spaces for wardrobes or spaces for cupboards to be built in later should be provided on the basis of 600 mm run of hanging space per person, with an internal depth of not less than 550 mm.

7.5 Bathrooms and W.Cs.

Detailed standards for the planning of bathrooms and W.Cs. are set out in Section 6.1.

COMMENTARY

7.4.1 The planning of double bedrooms should take particular account of the following:

(a) The space needed to:

    (i) Allow the furniture to be arranged in sensible alternative ways and to add the additional furniture required by (b).

    (ii) Avoid having beds alongside windows.

    (iii) Occasionally accommodate a cot in the main bedroom.

(b) Changing needs and flexibility in use: for example, double bedrooms, particularly when heated, may become:

    (i) Playrooms for small children.

    (ii) Studies and hobbies rooms for school-age children.

    (iii) Bed-sitting rooms for adolescent children and elderly relatives.

    (iv) Two separate rooms, particularly when children of opposite sexes reach adolescence.

(c) Built-in wardrobes: although more costly than simply providing space for the tenant's wardrobe, they can make a more compact and economical room layout possible.

(d) The design and positioning of doors, windows or ventilators: these should:

    (i) Conform with the appropriate Building Regulations.

    (ii) Minimise draughts.

    (iii) Ensure adequate lighting.

    (iv) Maintain privacy and sound insulation.

    (v) Be usable safely, without having to climb on furniture.

Diagrams 20-28 show the space needs of occupants for circulation and other activities in bedrooms.

7.4.2 The planning of single bedrooms should take particular account of the following:

(a) The space needed to:

    (i) Allow the furniture to be arranged in sensible alternative ways and add the additional furniture required by (b). See also Standards 7.4.3 and 7.4.4.

    (ii) Avoid having beds alongside windows.

(b) Changing needs and flexibility in use: when heated single bedrooms, as well as double bedrooms, may become:

    (i) Play rooms for small children.

    (ii) Studies and hobbies rooms for school-age children.

    (iii) Bed-sitting rooms for adolescent children and elderly relatives.

(c) Comments (c) and (d) on the planning of double bedrooms, which apply equally to single bedrooms.

Diagrams 27 and 28 give guidance on space requirements:

Note that compliance with a minimum apartment area is no longer required.

8

Diagrams

Diagrams 1-97 provide a wide range of dimensional data which will help the designer to evolve spaces suitable for the activities to be carried on in them.

The diagrams are arranged in three groups:

(A) Activities and Space Requirements

(B) Furniture sizes

(C) Equipment sizes

8(A) Activities and Space Requirements

Diagrams 1-42 show the space requirements of furniture and fittings and the adjacent spaces reasonably required by users. Some of the diagrams illustrate the space needs of activities by a group of people; others show the space needs for individual activity. The furniture and equipment sizes used are those shown separately in diagrams 43-97 in groups (B) and (C).

For convenience, the diagrams in group (A) are arranged under the following headings:

Food preparation
Meals space
Leisure space
Sleeping space
Personal care
Entrance and circulation space

Dimensions in this group have the following classifications:

(1) Overall sizes of furniture and equipment: as figured

(2) Distances between people or furniture and walls or
furniture higher than table-top: as figured

(3) Distances between people or furniture and
furniture of table-top height or less: figured and marked ■

(4) Distances between people or furniture and
furniture of seat height or less: figured and marked ▲

Note that the dimensions of spaces are for general guidance only and do not indicate absolute minima; nor should the room layouts shown be regarded necessarily as the only or the best ones for the activities in question.

FOOD PREPARATION

1. at the sink
passing with a tray
at the oven.

WORKTOP—COOKER
—WORKTOP—SINK—
WORKTOP,
uninterrupted by a door
or other traffic way.

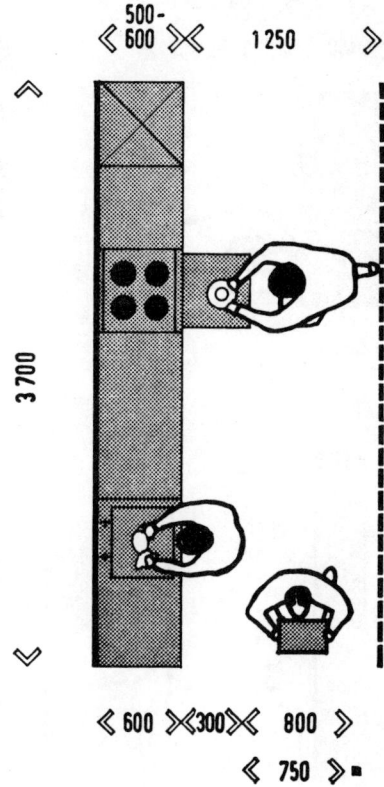

2. sitting at pull-out
worktop.

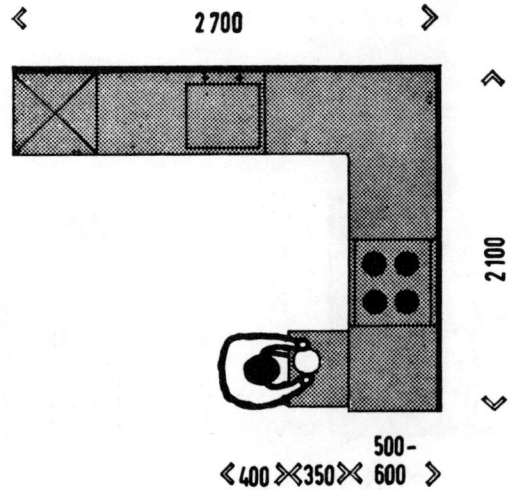

3. cooker rings and
worktop at the same
height.

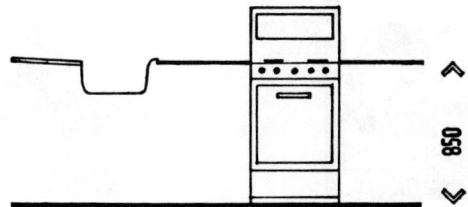

c

31

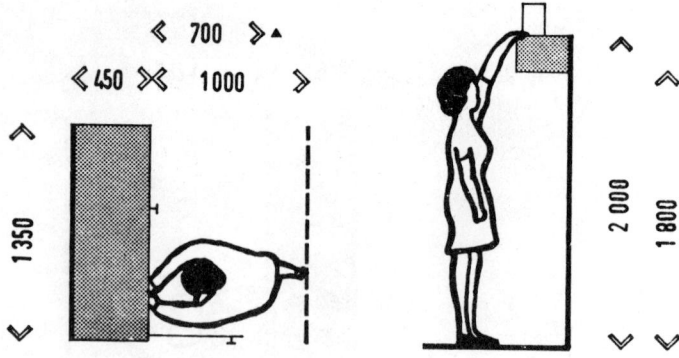

In diagrams 5-10 a woman of
average height is assumed
(1 630 mm).

4.   taking things from a
         low cupboard.

5.   maximum vertical reach
         and maximum shelf
         height for general use.

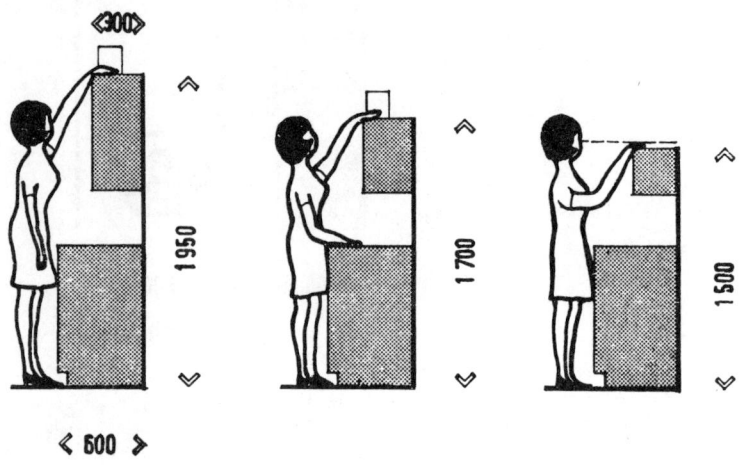

6.   maximum vertical reach
         over worktop.

7.   comfortable vertical
         reach over worktop.

8.   shelf at eye level.

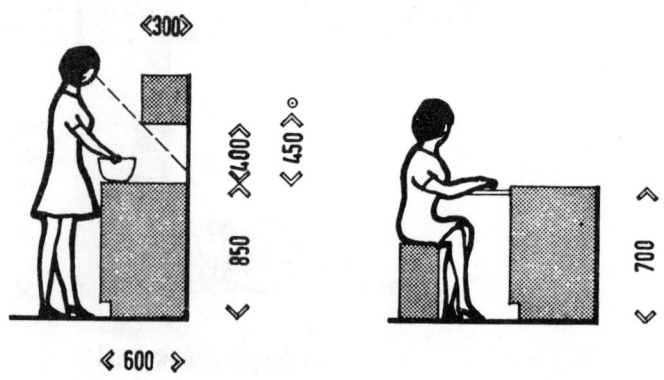

9.   comfortable height of
         worktop when standing.

10.   comfortable height of
         worktop when sitting.

TO BE AVOIDED

11.   cooker next to door.

12.   side-hinged cupboard
         doors at eye level.

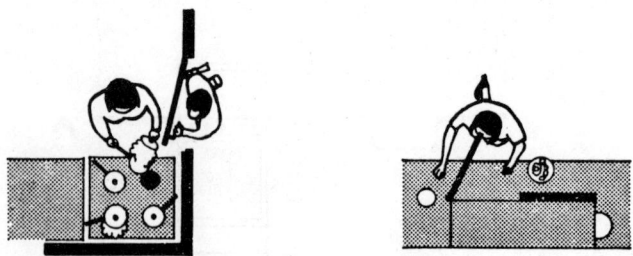

MEALS SPACES

13. space in kitchen for casual meals.

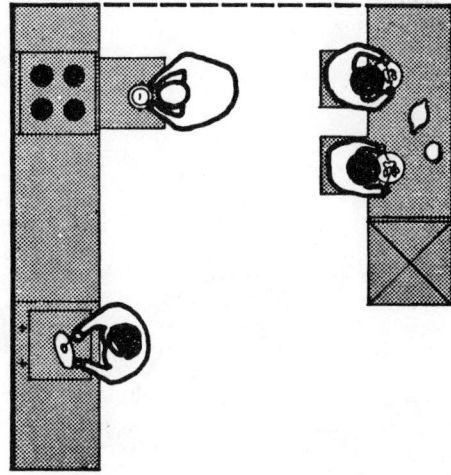

14. sitting at table and moving around.

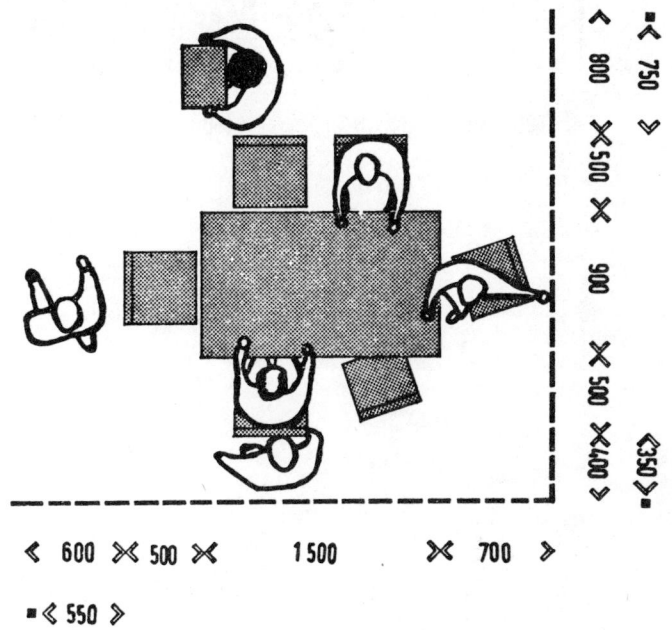

750
800
500
900
500
400 350

600   500   1 500   700

550

15. visitors entertained to meals.

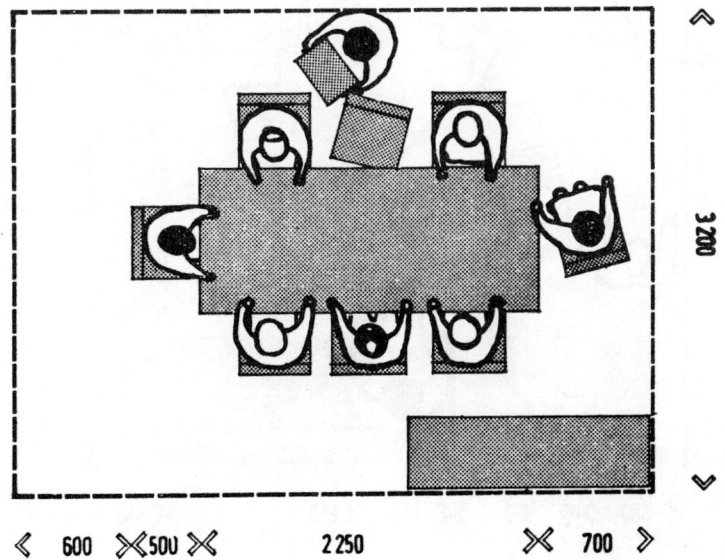

3 200

600   500   2 250   700

16. eating at a coffee table.

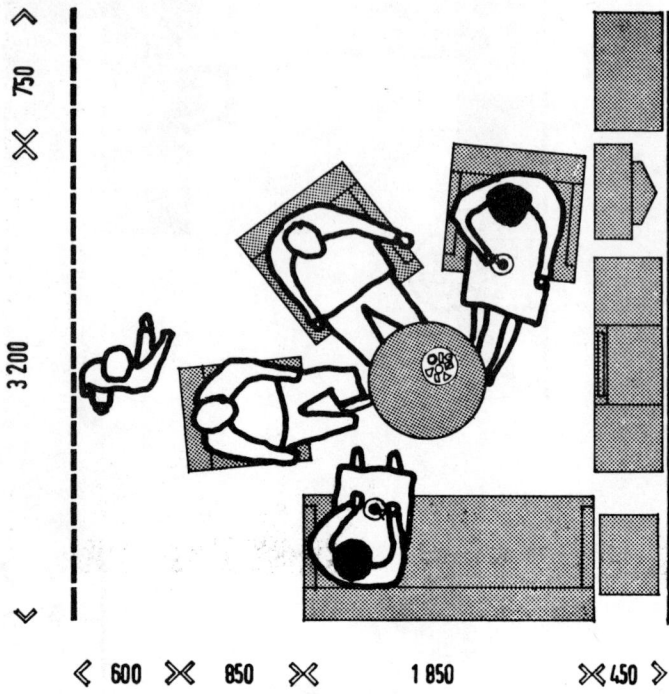

17. talking and reading.

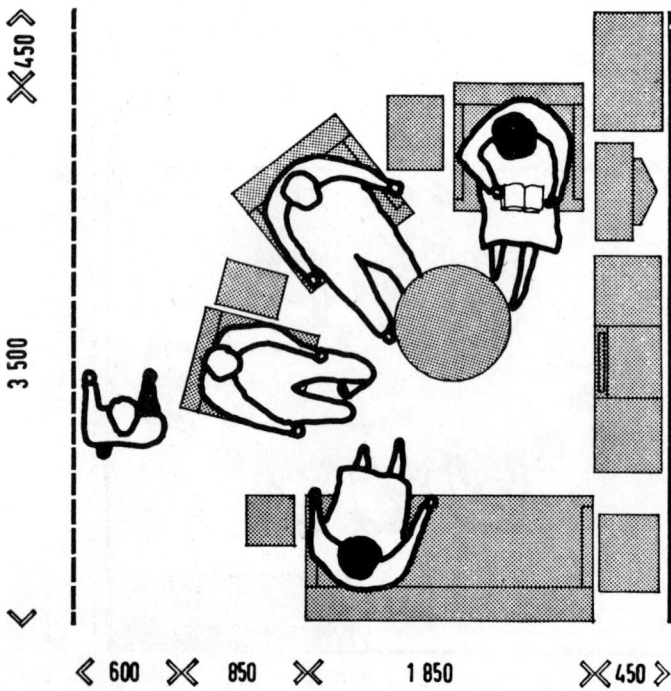

18.   looking at television.

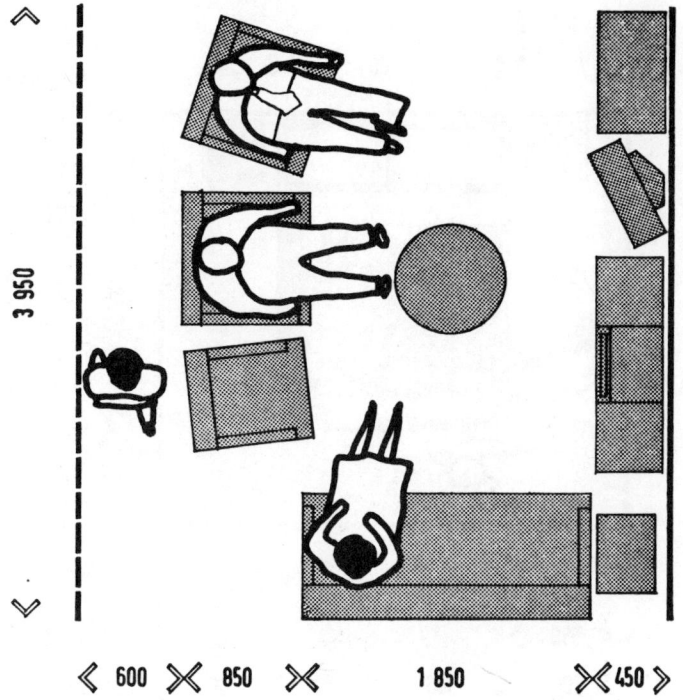

3 950

⟪ 600 ⨯ 850 ⨯ 1 850 ⨯ 450 ⟫

⟪ 850 ⨯ 1 850 ⨯ 450 ⟫

19.   entertaining visitors.

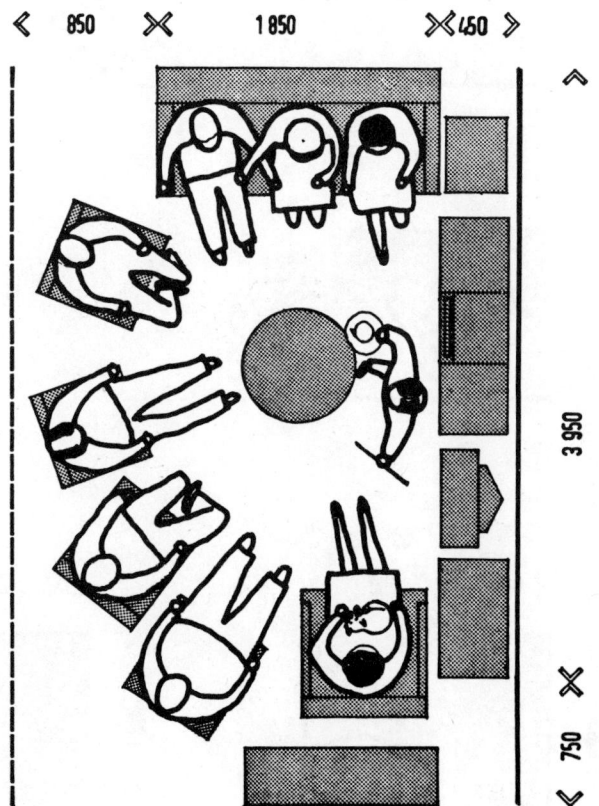

3 950

750

c•

650
700   1500   700

2 000

550   500

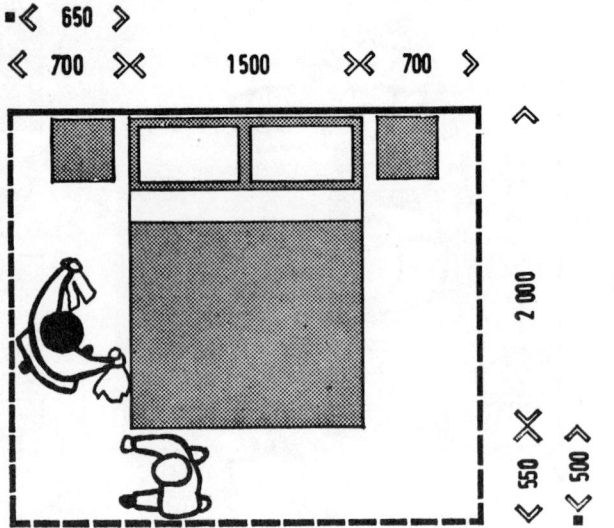

20.   circulation around a double bed.

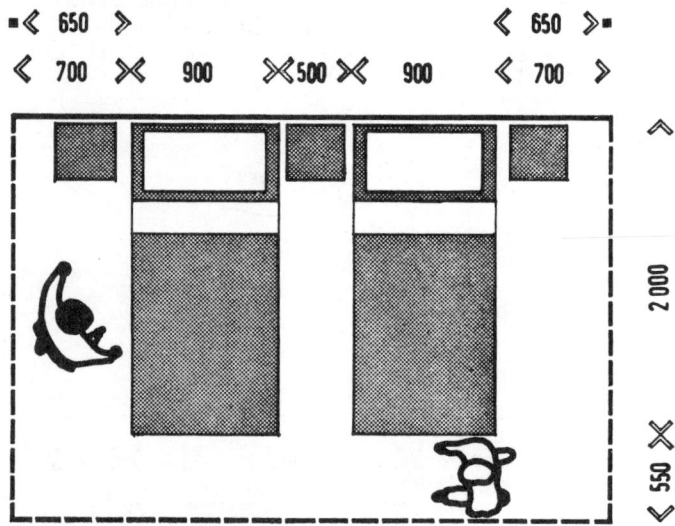

650   650
700   900   500   900   700

2 000

550

21.   circulation around twin beds.

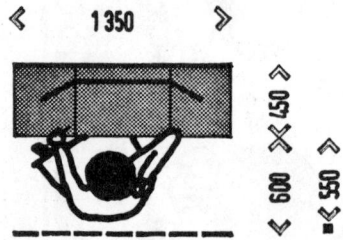

1 350

450

600   550

22.   at the dressing table.

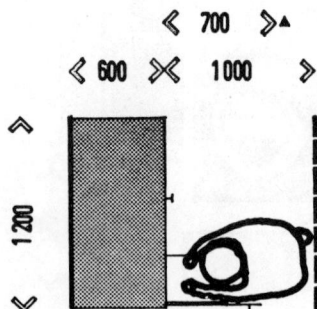

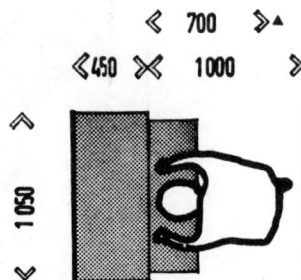

700
600   1000

1 200

700
450   1000

1 050

23.   opening a drawer

24.   at the chest of drawers.

25. twin beds may abut
when placed away from
walls.

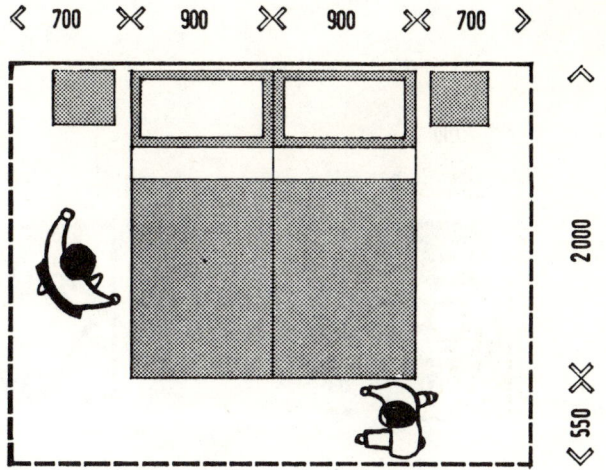

26. when twin beds are
placed alongside walls,
a space of 750 mm must
be left between them.

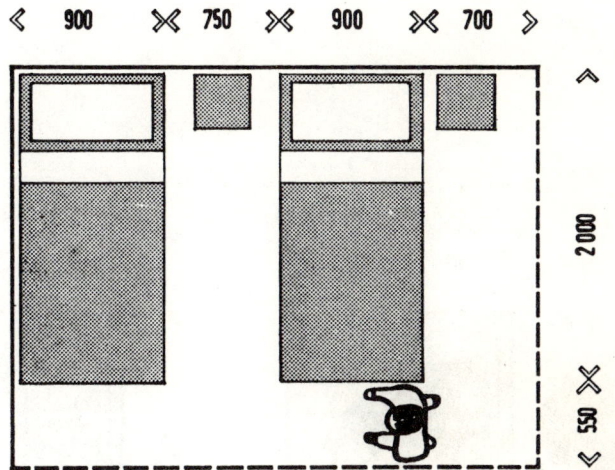

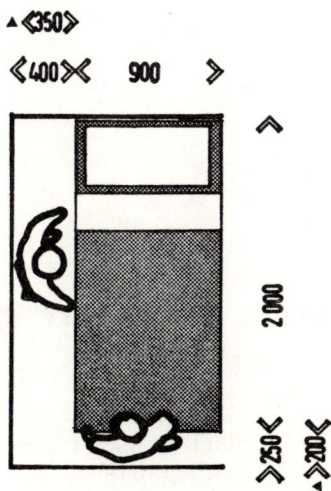

27. making a bed.

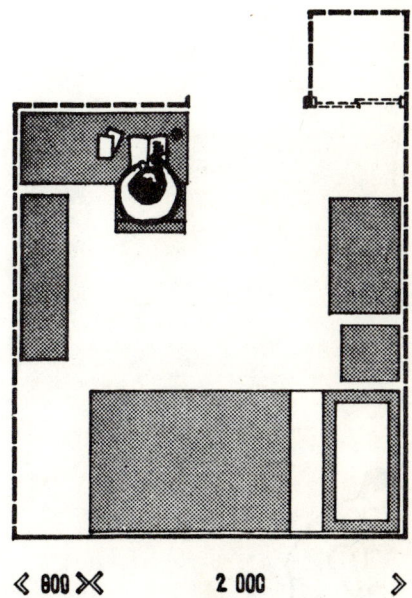

28. the use of a single
bedroom as a study
bedroom.

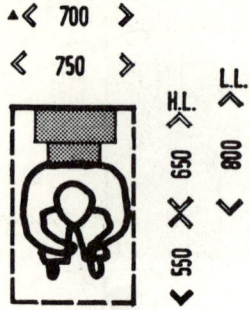

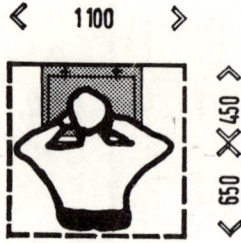

29.   using the W.C.

30.   using the wash hand
      basin.

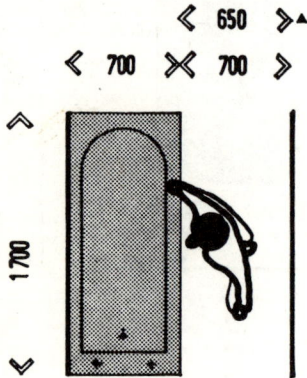

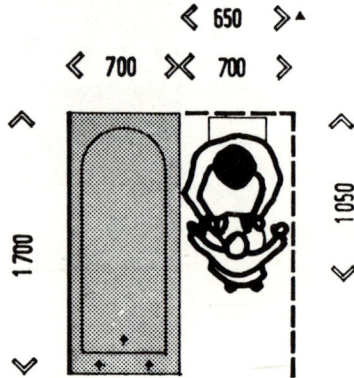

31.   drying oneself after
      a bath.

32.   drying a child after
      a bath.

DESIGN FOR SAFETY

33.   flat bottomed baths
      with grab rails.

CIRCULATING GENERALLY

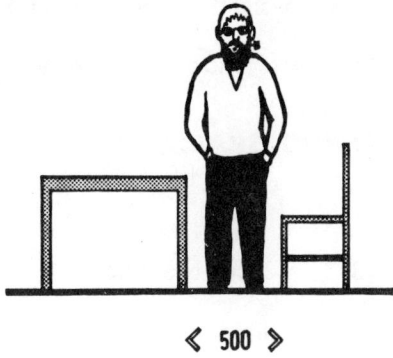

≪ 500 ≫

≪ 550 ≫

≪ 600 ≫

34.
35. passing between furniture of different heights.

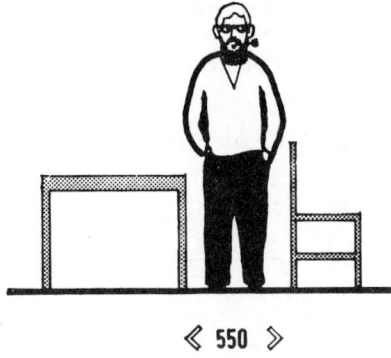

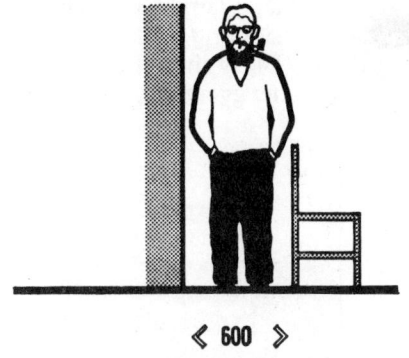

36. passing between furniture and wall.

HALLS AND LOBBIES

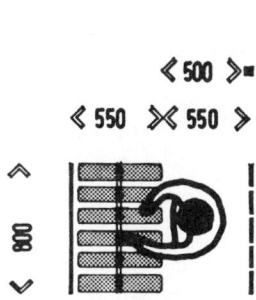

37. hanging coats on hangers.

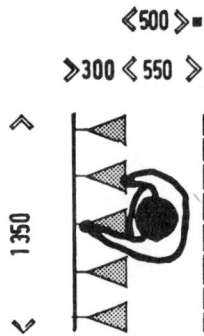

38. hanging coats on hooks.

39. helping on with a coat.

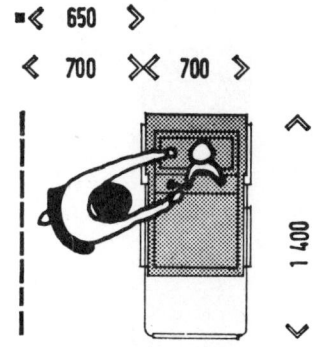

40. getting the pram ready.

STAIRCASES AND LANDINGS

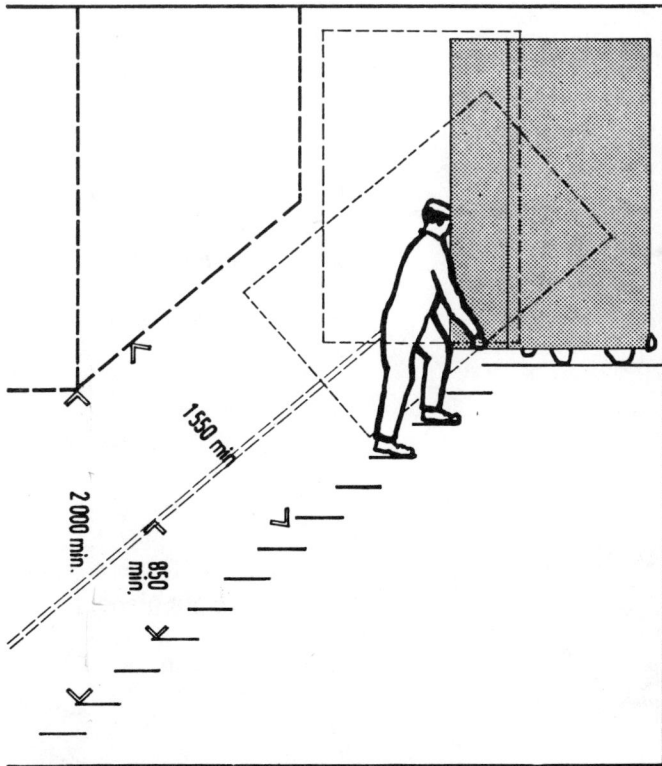

1550 min.

2000 min.

850 min.

41. moving a large wardrobe up a staircase showing minimum headroom and recommended handrail height.

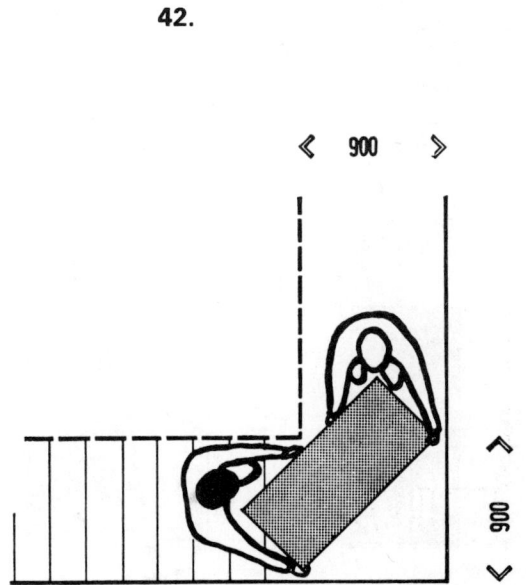

42.

≪ 900 ≫

900

8(B) Furniture Sizes

Diagrams 43-74 give dimensional data for a wide range of furniture, which will assist the designer in considering room layouts.
Sizes shown derive from an assessment of leading manufacturers' current products ; most should be taken as averages, although some are maximum.

LIVING SPACES

1:50

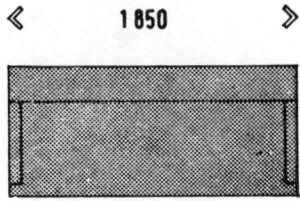
1 850 / 850

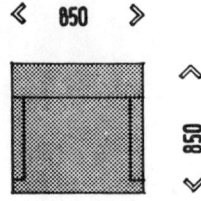
850 / 850

43.	3-seater settee.
44.	armchair.

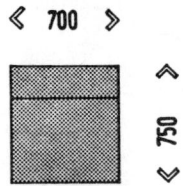
700 / 750

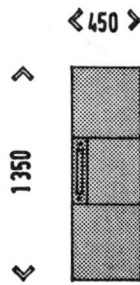
450 / 1 350

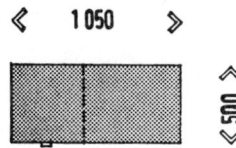
1 050 / 500

45.	easy chair.
46.	radiogram.
47.	desk.

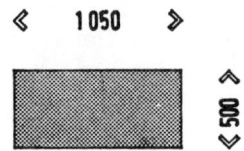
1 050 / 500

750 / 300

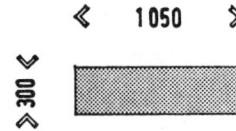
1 050

48.	coffee table.
49.	round coffee table.
50.	bookcase.

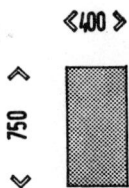
400 / 750

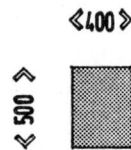
400 / 500

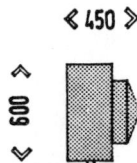
450 / 600

51.	record cabinet.
52.	nesting tables.
53.	television set.

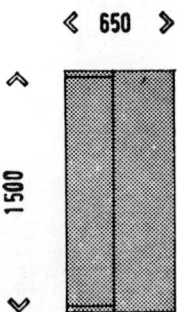
650 / 1 500

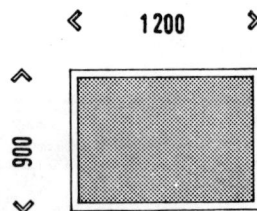
1 200 / 900

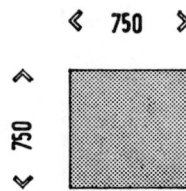
750 / 750

54.	upright piano.
55.	playpen.
56.	card table.

40

MEALS SPACES

57. dining table for 4.
58. dining table for 6.

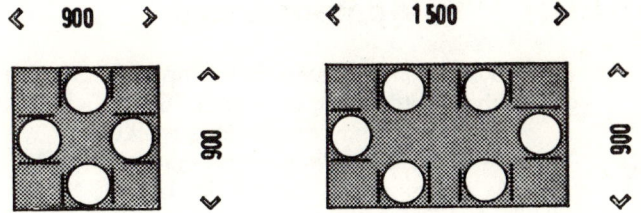

59. dining chair.
60. sideboard.

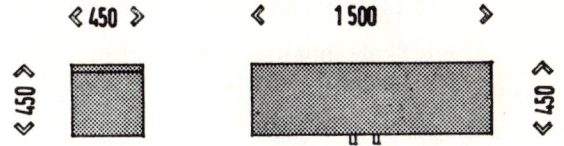

61. trolley.
62. stool.
63. high chair.

BEDROOMS

64. double bed.
65. single bed.
66. bedside table.

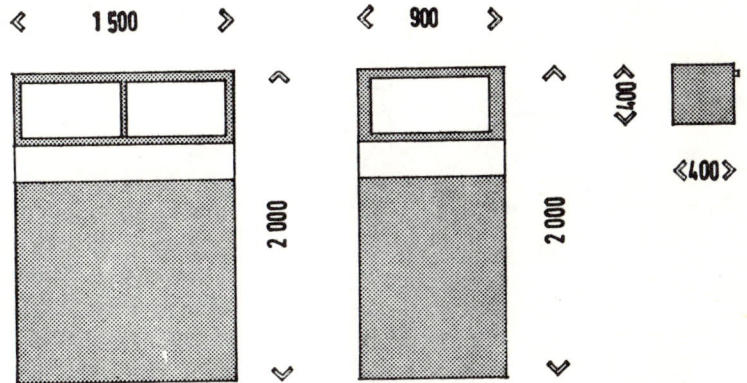

67. cot.
68. carry cot.

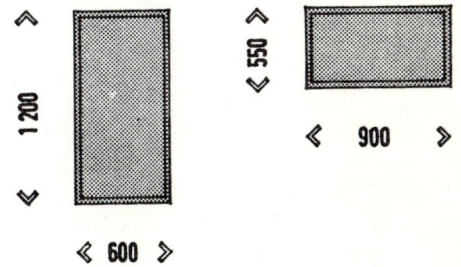

69. wardrobe—small.
70. wardrobe—medium.
71. wardrobe—large.

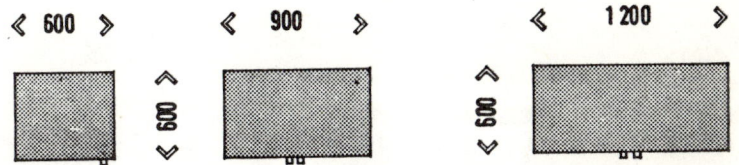

72. chest of drawers (small)
73. chest of drawers (large).
74. dressing table.

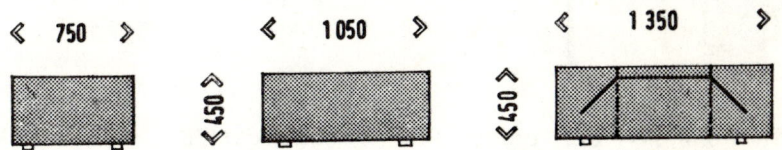

8(C)   Equipment Sizes

The dimensions given, in multiples of 50 mm, illustrate the range of current market sizes. In the case of kitchen equipment, diagrams 75-80, heights, widths and depths are shown with

minimum dimensions marked ○
average dimensions marked ⊙
maximum dimensions marked ●

The dimensions chosen will depend upon the size of the dwelling and upon which items of equipment are to be built in and which to be provided by the occupier.

It is likely that the dimensions of kitchen equipment will change when new metric standards are published by the British Standards Institution.

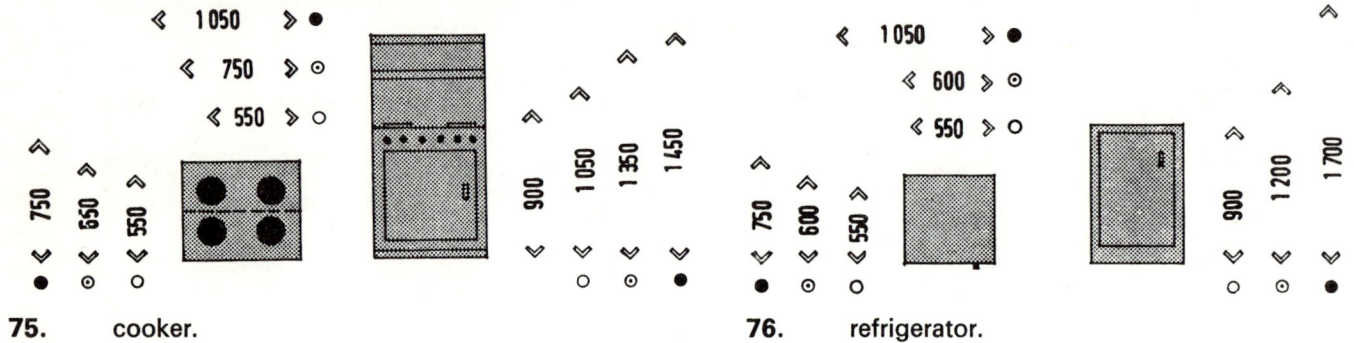

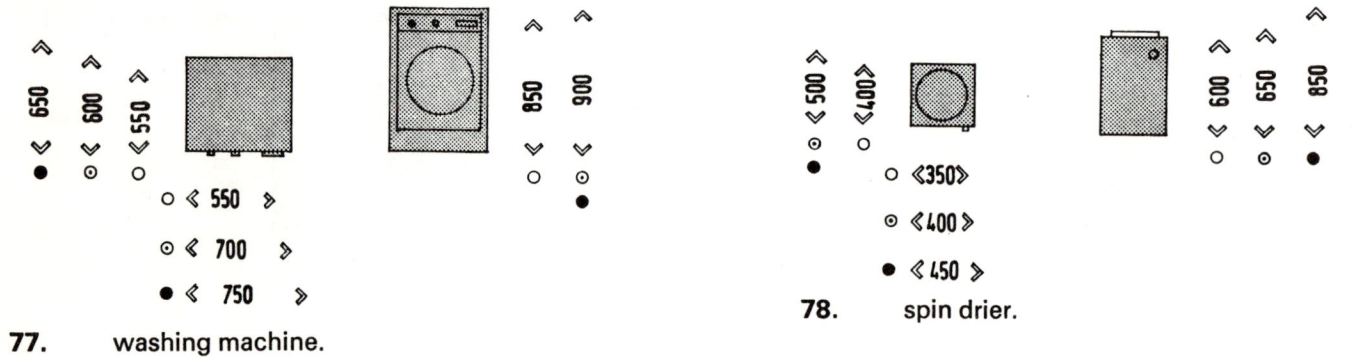

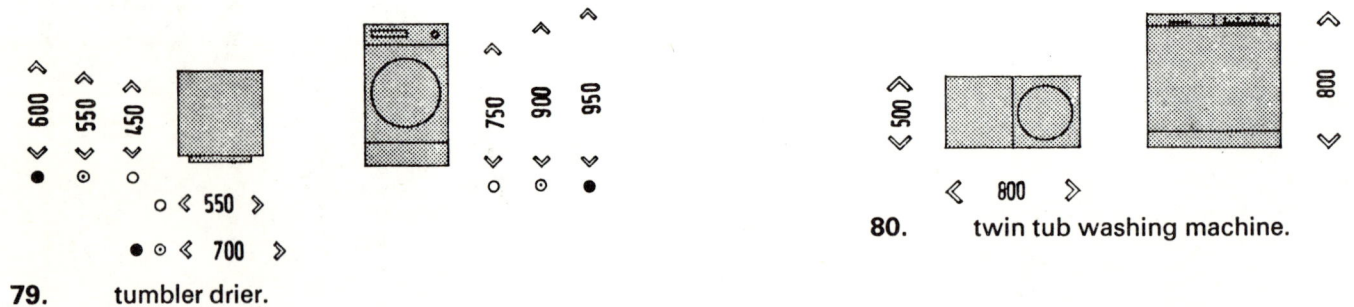

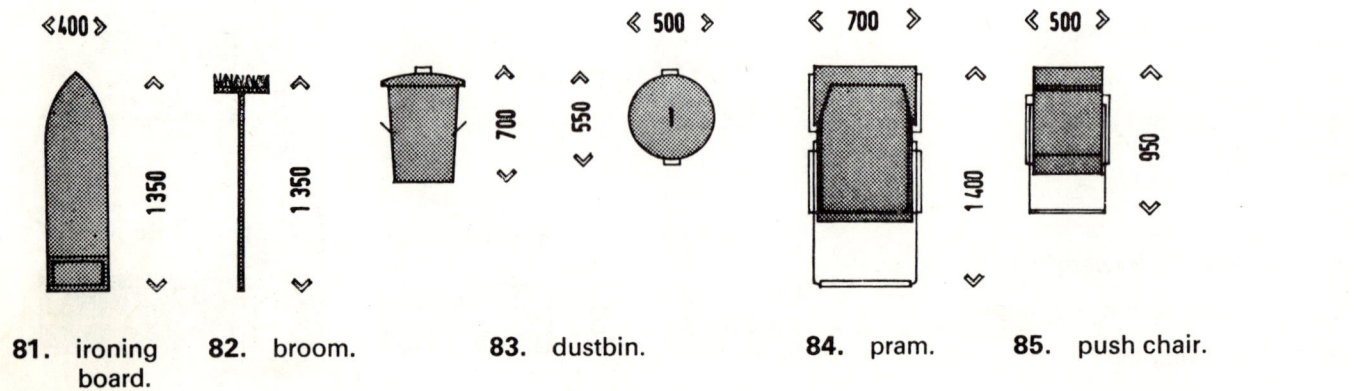

75.   cooker.

76.   refrigerator.

77.   washing machine.

78.   spin drier.

79.   tumbler drier.

80.   twin tub washing machine.

81.   ironing board.

82.   broom.

83.   dustbin.

84.   pram.

85.   push chair.

86. wheel barrow.
87. lawn mower.
88. garden roller.

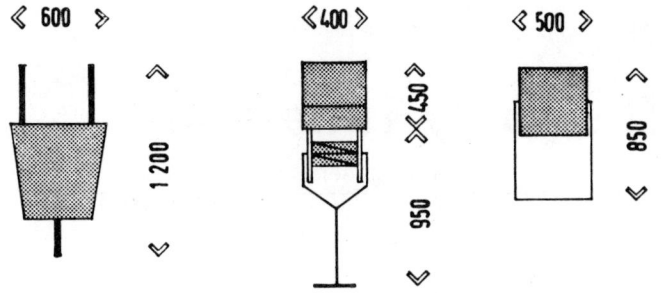

89. deck chair.
90. swing.
91. tricycle.

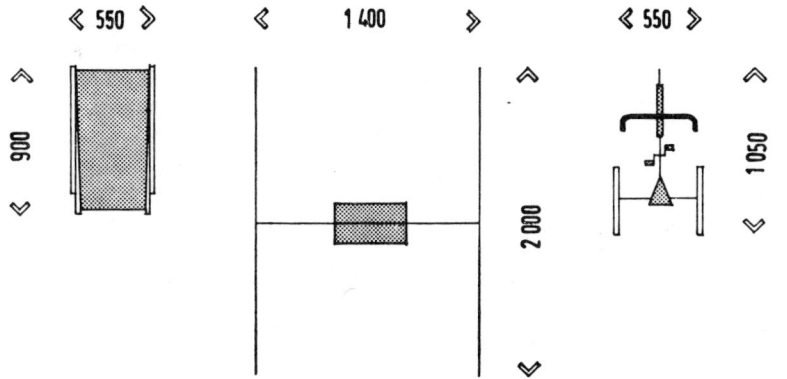

92. pedal car.
93. bicycle : handlebar
    height
    1 050 mm.
94. child's scooter.

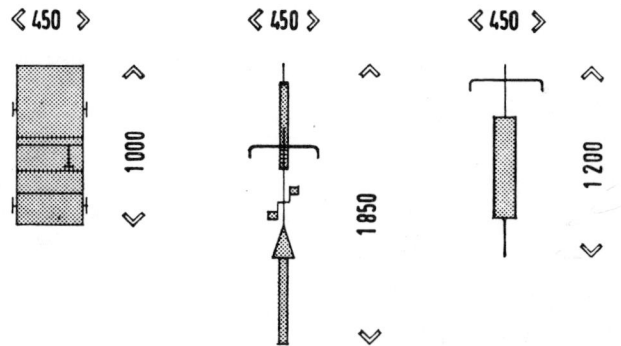

95. motor scooter.
96. motor-cycle.
97. motor car.
   (approximately 95% of
   cars on the British market
   have dimensions of less than
   4·78 m. long × 1·78 m wide
   × 1·83 m. high)

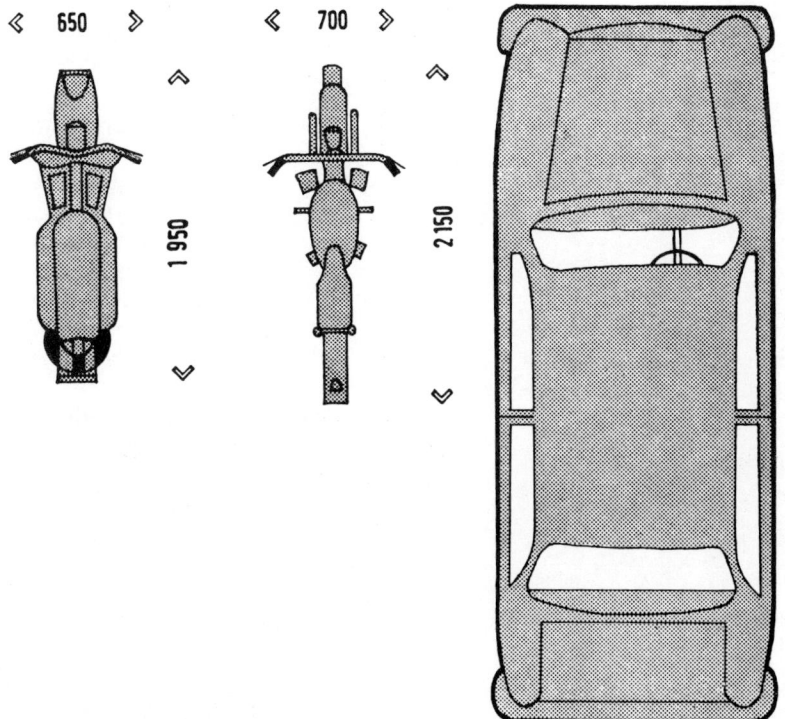

9
The Daily Timetable

9.1 THE RANGE OF HOME ACTIVITIES

The design of a dwelling has to cater for an enormous number of activities, and the designer should start from an awareness of the main ones. For example, the members of a family are continually coming and going, and callers of all kinds come to the home. The housewife is for much of her day engaged in some sort of housework—cleaning, washing ironing, cooking, clearing up after meals—and looking after children. The family come together and separate for different activities. Formal meals, television, entertaining friends, usually bring them all into one room; homework, reading or pursuing a hobby, tend to disperse them. Personal care and hygiene and activities like home decorating or gardening or cleaning the car also need to be considered.

These references to activities provide a starting-point from which designers can work forward, developing their own assessments of the full range of activities likely to be encountered in the different types of household for whom they may be designing. At the same time it must not be forgotten that a home is more than a place in which people do things. They also want to rest and relax, to enjoy their leisure time and bring up their children in attractive and comfortable surroundings.

9.2 THE GROUPING OF ACTIVITIES

As designers evolve their plans against a background of functional requirements so they must assess how activities are likely to be grouped together within the frame-work of a particular house plan. This will reveal how far the design will allow activities to be conveniently grouped in ways likely to be preferred by those living in the dwelling, and it will also indicate some of the alternative plan arrangements. Compromises will nearly always be unavoidable, especially in the smaller home, but care should be taken to ensure that the compromise chosen does the least violence to the expected grouping of activities.

When considering how activities can be grouped in different spaces or rooms in a dwelling, they can usefully be divided into the categories of primary and occasional. No rigid demarcation line between the two can be drawn, and how the activites are fitted into the available spaces, whether as primary or occasional, will depend very much on the way in which the total living space is divided up. But usually, certain activities take place mainly in a particular room or area where they can be treated as primary, and around them cluster a wide variety of activities which take place only occasionally.

The primary activities in the living area generally engage several members of the family, and therefore tend to make the living room a 'noisy area'. Looking at television is an obvious example of this. Many of the occasional activities will be quieter, such as homework or sewing. As some of them will not coincide in time with primary activities, the living area can often be used without difficulty for both. But where 'noisy' and 'quiet' activities are likely to coincide, as for example looking at television, and study or homework during the evenings, the house plan should enable both kinds of activity to be carried on at the same time. This can be done in various ways. For example, if the plan allows for the dining space in a separate room, this may also be used for quiet activities. Alternatively, if more convenient, noisy activities could be transferred to the dining space, thus freeing the living space for quiet activities. Another possibility, if the dining space and kitchen are combined, is to ensure that some bedroom space is suitable for quiet activities. This has various design implications both for how and where the bedrooms are arranged, and for heating and the arrangement of furniture.

Apart from trying to assess which activities should be treated as primary and occasional or as noisy and quiet in particular cases, it is also necessary to ask whether they are 'tidy' or 'untidy'. The separation of 'tidy' from 'untidy' is always difficult and more so in a small dwelling. Practically all activities create some untidiness, especially having meals and looking after young children. If there is a small but well-arranged working kitchen, some untidy activities will have to be carried on in the dining space or living room, e.g. children playing. If the dining space is separate (or separable) from the living room, untidy activities of this kind can be concentrated in one of them. On the other hand, the provision of a larger kitchen might make it possible for the children to play there under their mother's eye.

The points which have just been put forward are merely examples of the likely groupings of activities which the designer must take into account. There are others he will think of for himself according to the type of household, local habits and the pattern of living.

9.3 TIME AND PLACE OF ACTIVITIES

The two sets of diagrams in Figs. A and B illustrate, for a family at two different stages in its development, how activities coincide in the home at successive periods of the day. They also give an indication of 'place' as well as showing a pattern of movement about the dwelling. They show the separating out of individual members, and the coming together of the family as a whole. Together they indicate particular points of stress, when the home is being most intensively used. The exact nature of these stresses, and of the patterns of movement, will vary with the way of life of the household, but this kind of analysis based on local knowledge can be of immense help to the designer.

TIME AND PLACE OF ACTIVITIES

Fig. A. The Younger Family

(Parents and three children: a boy of school age (7) and girls of 3 and 1)

0700   In the early morning rush, instant hot water and warmth are needed.

0710   Breakfast has to be served quickly, the school child got ready and the other children looked after as they wake up.

0830   Father and school child are off. Mother gives the other children their food and has something herself. A place where food can be eaten near the work area is useful.

0930   Mother puts the baby out in the pram and the toddler plays outside. Mother needs to be able to see the children easily while she works.

1130   Returning from shopping, Mother needs space to put the pram and the shopping and room to take off the children's outdoor clothes, and somewhere convenient to put them.

1200   When the children play indoors Mother needs to be able to see them from the kitchen, but they should be away from the kitchen equipment and not under her feet.

1230   If the family comes home for a midday meal on week days they have to wash and eat quickly. The dining space should be conveniently reached from the work centre.

1430   The baby needs a place where it is quiet to sleep. The toddler needs a place for play where toys can be concentrated and Mother does not have to be for ever tidying up.

46

1530 Space in the tidy area of the house is needed for visitors, while the children of both families play.

1700 Watching television is a major family activity with the children now in the living room.

1800 The evening meal may be taken while watching television. Space for a low table is needed.

1830 People do not always wish to watch television, and may need space to sit away from it. The children need quiet as they settle to sleep.

1900 When Father makes or repairs something, he should not be in Mother's way in the kitchen, nor where he will disturb sleeping children.

2000 Sometimes visitors are being entertained while a child watches television.

2200 Mother may want to talk to visitors while she is preparing some snacks for them.

2330 The parents need to sleep near their young children, so that they can attend to them easily.

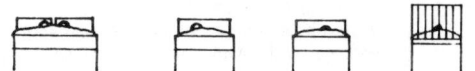

0300

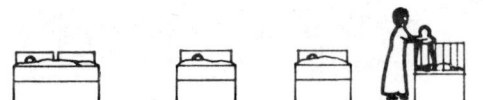

Fig. B. The Older Family

(Parents (mother working part-time) and boy aged 23, girl 20, boy 14)

0700   With 4 workers and a secondary school child wanting to wash before they leave home, a second W.C. and wash basin are needed. Hot water and warmth are again essential.

0730   There is a crush in and around the kitchen as breakfast is eaten and other preparations made before leaving the home.

0830 to 1630   While the house is empty during the day, the bread, milk, parcels and perhaps laundry may have to be delivered and put in safe places, and the meter reader may call.

1630   When the wife returns from work she wants to be able to warm the house, clear up and get a snack with as much speed and as little trouble as possible.

1830   The evening meal may be the only time during the week when the family sit down together. They may like to eat away from the kitchen area.

2000   In a practically adult family several individual activities may take place in an evening at home and space is needed for them.

2100   The family will sometimes split up into groups during an evening and the children may entertain their own friends separately. Room and privacy are needed for more than one group.

2230   Before going to bed, people at work often have to get things ready for the morning and meantime perhaps have a snack. Space is needed for several people to do their chores at once.

2330   Separate bedrooms are needed by each child when reaching adolescence, but they do not need to be near the parents.

48

10

References : Bibliography

10.1   References

10.1.1   CROSS-REFERENCES

The following table notes against the new standards those of the Building Regulations and paragraphs from the Scottish Housing Handbook Part 3 which refer:

New Space Standards	Relevant Bdg. Regs.	Relevant Paras: from SHH 3
4.1 and 4.2	183, Schedule 1.	7, 16, 23(d), App. A, B
4.4.1	189	23(d), App. A
4.4.2	188	23(b), App. A
4.4.3	189	23(c), App. A
4.4.4 (see 6.2)	See 6.2	See 6.2
5.1	193, Part X	18, 27, App. C
6.1	116, 173, 185	22
6.2	114, 132, 186 187, 190, 191	20, 23(a), 23(d)
6.3	195	None
7.1	179, 182	7
7.2	None	24
7.3	None	19
7.4	None	21, 23(c)

10.2   Acknowledgement

The Scottish Development Department acknowledges the co-operation of the Ministry of Housing and Local Government in the preparation of this Bulletin.

10.3   Bibliography

The following HMSO publications are generally recommended:

'Homes for Today and Tomorrow'.	M.H.L.G.	1961	(4s. 0d. net.)
'Houses and People': a review of user studies at the Building Research Station.	Min.Tech.	1966	(27s. 6d. net.)

10.3.1

The following HMSO publications are recommended subject to compliance with the Building Regulations:

'Space in the Home': metric edition DB6.	M.H.L.G.	1968	(5s. 0d. net.)
'Safety in the Home': DB 13.	M.H.L.G.	1967	(4s. 6d. net.)
'House Planning': a guide to user needs, with check list: DB 14.	M.H.L.G.	1968	(8s. 6d. net.)
Co-ordination of Components in Housing: Metric Dimensional Framework: DB 16.	M.H.L.G.	1968	(6s. 0d. net.)

(Prices in brackets include postage)

50

Index

Printed in Scotland for Her Majesty's Stationery Office by Buccleuch Printers Ltd, Hawick
Dd. 762089/2397 C5 4/85